SOCIAL PERSPECTIVES OF AGING

THE GLOBAL CHALLENGE

AGING POPULATIONS, BIO-MEDICINE AND CHINA

SOCIAL PERSPECTIVES OF AGING

JASON L. POWELL - SERIES EDITOR –

DEAN OF FACULTY, UNIVERSITY OF LIVERPOOL, UK

Aging in Perspective and the Case of China: Issues and Approaches
Sheying Chen and Jason L. Powell (Editors)
2011. ISBN: 978-1-61728-941-5

Social Welfare, Personal Budgets and Care: A Case Study
Jason L. Powell
2011. ISBN: 978-1-61209-332-1 (Softcover)

Aging and Social Policy: A Foucauldian Excursion
Jason L. Powell
2011. ISBN: 978-1-61209-838-8

A Short Introduction to Social Theory
Jason L. Powell
2012. ISBN: 978-1-62100-928-3 (Softcover)

Understanding the Voluntary Sector: Critical Success Factors - A Case Study
Jason L. Powell
2012. ISBN: 978-1-62100-499-8 (Softcover)

The Global Dynamics of Aging
Jason L. Powell and Sheying Chen (Editors)
2012. ISBN: 978-1-62100-936-8

Exploring Policing, Crime and Society: A Brief but Critical Understanding
Jason L. Powell
2012. ISBN: 978-1-61942-001-4 (Softcover)

Foundations of Social Inquiry: A Brief Excursion
Jason L. Powell
2012. ISBN: 978-1-61942-004-5 (Softcover)

Foucault: Interdisciplinary Approaches
Jason L. Powell (Editor)
2012. ISBN: 978-1-61942-146-2

Aging, Gender and Crime
Jason L. Powell
2012. ISBN: 978-1-61942-150-9 (Softcover)

Aging, Theory and Globalization
Jason L. Powell
2013. ISBN: 978-1-61761-947-2 (Softcover)

The Sociology of Care: Exploring Theory, Policy and Practice – The Case of Aging
Jason L. Powell
2013. ISBN: 978-1-62100-697-8
(Softcover)

China, Aging and Theory
Jason L. Powell
2013. ISBN: 978-1-62100-915-3
(Softcover)

Understanding Aging: A Short Tour
Jason L. Powell
2013. ISBN: 978-1-62417-196-3
(Softcover)

Volunteering, Aging and Business
Jason L. Powell
2013. ISBN: 978-1-62618-743-6
(Softcover)

Globalization and Global Aging
Jason L. Powell
2013. ISBN: 978-1-62618-744-3
(Softcover)

Social Gerontology
Jason L. Powell
2013. ISBN: 978-1-62618-745-0
(Softcover)

Later Life
Jason L. Powell
2013. ISBN: 978-1-62618-746-7
(Softcover)

SOCIAL PERSPECTIVES OF AGING

THE GLOBAL CHALLENGE

AGING POPULATIONS, BIO-MEDICINE AND CHINA

JASON L. POWELL

nova publishers
New York

For permission to use material from this book please contact us:
Telephone 631-231-7269; Fax 631-231-8175
Web Site: http://www.novapublishers.com

Library of Congress Cataloging-in-Publication Data

The global challenge : aging populations, bio-medicine and China / editor, Jason L. Powell (University of Central Lancashire, UK).
pages cm
Includes index.
ISBN 978-1-62808-214-2 (soft cover)
1. Older people--China. 2. Aging--China. 3. Biomedical engineering--China. I. Powell, Jason L., 1971-
HQ1064.C6G56 2013 305.260951--dc23 2013020305

Published by Nova Science Publishers, Inc. † New York

To the memory of Elizabeth Joyce Hawkins

Her influence shines through on every single page

CONTENTS

PREFACE

This book explores the global challenge: understanding aging as a medical concept, social concept and population term. The focus is on China and global aging. The book further assesses the rapid rise of aging populations what could be entitled 'The Silver Tsunami' in highlighting how an unstoppable force of populational aging needs to be analysed what the social and economic implications are for all continents across the globe.

Chapter 1

INTRODUCTION

Economists and policy makers around the globe are increasingly concerned about the rising numbers of older people in their society. There are worries about the inadequacy of pension funds, of growing pressures on welfare systems, and on the inability of shrinking numbers of younger people to carry the burden of their elders.

This book focuses on such issues in China, where the older people have become a rapidly expanding proportion of the population. While resources do need to be targeted on the vulnerable older people, the presumption that older people as a whole are an economic and social burden must be questioned. This is an ageist view that needs to be combated by locating how bio-medical views on aging seep into policy spaces in China that position negative perceptions of aging as both individual and populational problems.

The first chapter focuses on this and then moves to observe the implications of bio-medicine for older people in China in terms of "vulnerable" aging but deconstruct such "fixed" explanations by juxtaposing *active aging* as key narrative that epitomizes "declining to decline" as espoused by bio-medical sciences. The following chapter explores the emergence of social theories of aging that need to be considered as a counter-point to understanding aging in China and across the globe. The final chapter explores the rapid rise of aging populations what could be entitled “The Silver Tsunami” in highlighting how an unstoppable force of populational aging needs to be analysed what the social and economic implications are for all continents across the globe.

Chapter 2

TAKING AIM AT OLDER PEOPLE IN ASIA: BIO-MEDICINE

There is no doubt that in many societies around the globe older people are a growing proportion of the global population (Krug 2002). The age structure of the global population is changing from one in which younger people predominated to a society in which people in later life constitute a substantial proportion of the total population. While the biological and psychological models of aging inscribe it as an "inevitable" and "universal" process, an aging population is neither (Phillipson, 1998). Transformations in the age profile of a population are a response to political and economic structures. A major concern for organizations such as the United Nations and World Bank focuses on the number of such "dependent" older people in world society (Krug 2002).

Indeed, older people in particular constitute a large section of populations in global aging. In relation to public services that have to be paid for by "younger" working people, the percentage of the population has been used to signify such "burdensome" numbers. Not only are older people seen as dependent but also children under school leaving age and people over the retirement age. Dependency rates—that is, the number of dependants related to those of working age—altered little over the twentieth century and yet the notion of "burden" group retains its legitimacy. The reason there has been so little change during a period of so-called rapid aging populations is that there has been a fall in the total fertility rate (the average number of children that would be born to each woman if the current age-specific birth rates persisted throughout her child-bearing life) (Phillipson, 1998).

Changes in the age structures of all societies also affect total levels of labor force participation in society, because the likelihood that an individual will be in the labor force varies systematically by age. Concurrently, global population aging is projected to lead to lower proportions of the population in the labor force in highly industrialized nations, threatening both productivity and the ability to support an aging population (Powell, 2009). Coupled with rapid growth in the young adult population in Third World countries, the World Bank (1994) foresees growing "threats" to international stability, pitting different demographic-economic regions against one another. That the United Nations (2002) views the relationship between aging populations and labor force participation with panic recognizes important policy challenges, including the need to reverse recent trends toward decreasing labor force participation of workers in late-middle and old age despite mandatory retirement in both Western and Eastern countries such as the UK (Jackson and Powell, 2001) and China (Chen and Powell, 2011).

Notwithstanding this, in China there is also an ongoing huge increase in the aging population that replicates global trends. It can be seen that the percentage of the aging population has increased from just over 4.4 percent to just under 7 percent from 1953 to 2000 (Cook and Murray 2001). Note that the increase has not been constant, reflecting the negative impact of the Great Leap Forward and the successive famines upon the demographic profile, shown in the results of the 1964 Census.

Since that latter date, the percentage of the elderly has nearly doubled, while the actual numbers have more than tripled, being approximately 3.6 times that of 1964 by the year 2000. By 2000, there were 88 million Chinese aged 65 or over, compared to just under 25 million in 1964, an increase of 63 million plus in 36 years.

As is well known, China's population policy, most usually referred to as the Single Child Family Program (although some Chinese commentators regard the phrase as a misnomer, given that more than one child is possible within a number of situations), has led to a rapid deceleration in the birth rate, which was only 13.38 per thousand in 2001, compared to 21.06 in 1990 and 18.25 in 1978 (Cook and Murray 2001).

This controversial policy, lambasted and praised in equal measure depending on perspective, has meant a rapid expansion in the number of single children only households at the very same time that the proportion of the elderly has also increased as a new era of prosperity has reached many households in China.

Estimates suggest that up to 300 million less people have been born as a result of this process of state intervention, but Murray (2004) has built on the earlier work of D.G. Johnson and others to suggest that improved living standards via modernization would have led to the same outcome voluntarily as growing numbers of urban dwellers in particular chose to reduce their family size. This last point hints at the important spatial dimension of demographic change in China. There is on the one hand a marked contrast between urban and rural life in China and, closely related to this, a marked contrast between Eastern China and Western China (Cook and Powell, 2007). It is in the heavily urbanized "Gold Coast" of the Eastern Seaboard in which China's spatial transformation is most dramatic, with fast-expanding cities being especially concentrated in the Pearl River delta of the Southeast and the Yangtse Delta in Central East China (Cook and Murray 2001).

Hence, it is not simply that the Chinese government has belatedly recognized "the greying" of populational constructions and policy implications, it is that they continue to look for knowledge of aging as the power to define old age as a social problem in terms of dualistic distinctions between deviancy and normality. An aging population, like that of an individual being studied by bio-medical models, is seen as a "burden" problem in terms of economic management of Eastern (and Western) economies.

It could be argued when looking at the effects of a so-called "demographic time bomb" across US, Europe, and Asia, it may have been grossly exaggerated.

Such old age, therefore, has been perceived negatively via a process of "ageism" — stereotyping older people simply because of their chronological age. Agist stereotypes such as "aging populations" act to stigmatize and consequently marginalize older people and differentiate them from groups across the life-course who are not labeled "old" (Bytheway, 1993).

One of the ways to interpret social aging such as being a categorization whether it be in individual or populational terms is through use of theorizing on what it means to age in society; that is, concerns and social issues associated with aging and the ways in which these themes are influenced and at the same time influence the society in which people live. Thus, to understand the process of aging, looking through the lens of the "sociological imagination" is not to see it as an individualized problem rather as a societal issue that is faced by both First World and Third World nations as a whole.

In supporting this latter view, there is a need to focus on how populational discourses of aging in China are influenced and reinforced by bio-medical models of aging that help drive perceptions of older people as a burdensome group.

WHERE DOES THE AMBIVALENCE TO AGING COME FROM?

There are important implications for how aging is viewed by not only in terms of global aging but more specifically to China and the arrangement of political and economic structures that create and sanction social policies grounded in knowledge bases of "burdensome" populations (cf Powell, 2001). Such knowledge bases are focused on: one, "biological aging" which refers to the internal and external physiological changes that take place in the individual body; two, psychological aging is understood as the developmental changes in mental functioning — emotional and cognitive capacities.

Bio-medical theories of aging can be distinguished from social construction of aging: (1) focusing on the bio-psychological constituent of aging, and (2) on how aging has been socially constructed. One perspective is driven from "within" and privileges the expression from inner to outer worlds. The other is much more concerned with the power of external structures that shape individuality. In essence, this social constructionism poses the problem from the perspective of an observer looking in, whilst the biomedical model takes the stance of inside the individual looking out (Powell and Biggs, 2000). There has long been a tendency in matters of aging and old age to reduce the social experience of aging to its biological dimension from which are derived a set of normative "stages" which over-determine the experience of aging.

Accordingly being "old," for example, would primarily be an individualized experience of adaptation to inevitable physical and mental decline and of preparation for death. The paradox of course is that the homogenizing of the experience of old age which the reliance on the biological dimension of old age entails is in fact one of the key elements of the dominant discourse on aging and old age.

It is interesting that comparative historical research on aging in Eastern culture highlights an alternative perception of aging; in 18th Century China has highlighted a rather different path as to the conceptualization of aging as a scientific process developed by western rationality. For example, Cook and Powell (2003) observe that traditional Chinese society placed older people on a pedestal. They were valued for their accumulated knowledge, their position within the extended family, and the sense of history and identity that they helped the family to develop (Murray,1998). Respect for elderly people was an integral part of Confucian doctrine, especially for the family patriarch.

This was a view that was also prevalent in Ancient Greece with the notions of

> "respect" for older people especially regarding gendered issues of patriarchy (cf Bytheway, 1993). Prior to industrialization, in India, there was a bestowment that older people had responsible leadership roles and powerful decision-making positions because of their vast "experience," "wisdom" and "knowledge" (Katz, 1996).

It seems with the instigation of Western science and rationality, aging began to be viewed in a different more problematic context than to the Confucian doctrine of aging epitomized in China and issues of respect for aging in India. Martin Heidegger (1971) makes the similar point when he spoke of the Westernization of the World through the principles of Western science and language.

Indeed, the technological developments due to industrialization, Westernization, and urbanization—under the purview of distorted form of modernity—have neglected these statuses of aging by downgrading its conceptualization. Part of understanding individuality in Western culture, the birth of "science" gave legitimate credibility to a range of bio-medical disciplines of whom were part of its umbrella. In particular, the bio-medical model has become one of the most controversial yet powerful of both disciplines and practice with regard to aging (Powell and Biggs, 2000).

The bio-medical model represents the contested terrain of decisions reflecting both normative claims and technological possibilities. Bio-medicine refers to medical techniques that privilege a biological and psychological understanding of the human condition and rely upon "scientific assumptions" that position attitudes to aging in society for their existence and practice. Hence, scientific medicine is based on the biological and psychological sciences. Some doctrines of the biomedical model more closely reflect the basic sciences while others refer to the primary concern of medicine, namely diseases located in the human body. Most important is that these beliefs hold together, thereby reinforcing one another and forming a coherent orientation toward the mind and body. Indeed, the mind-body dualism had become the location of regimen and control for emergence of scientific in a positivist methodological search for objective "truth." The end product of this process in the West is the "biomedical model."

In this sense, bio-medicine is based on the biological and psychological sciences. Some doctrines of the biomedical model more closely reflect the basic sciences while others refer to the primary concern of medicine, namely diseases located in the human body. Most important is that these beliefs hold together, thereby reinforcing one another and forming a coherent whole (Powell, 2001). By developping an all-encompassing range of bio-medical discourses, many forms of social injustice could be justified as "natural," inevitable and necessary for the successful equilibrium of the social whole such as mandatory retirement and allocation of pensions (Phillipson, 1998).

Bio-medical gerontology is a fundamental domain where medical discourses on aging have become located and this is very powerful in articulating "truths" about aging. Under the guise of science and its perceived tenets of value-freedom, objectivity and precision (Biggs, 1993), bio-medical gerontology has a cloth of legitimacy. Biological and psychological characteristics associated with aging have been used to construct scientific representations of aging in modern society. The characteristics of biological aging as associated with loss of skin elasticity, wrinkled skin, hair loss or physical frailty perpetuates powerful assumptions that help facilitate attitudes and perceptions of aging. It may be argued that rather than provide a scientific explanation of aging, such an approach homogenizes the experiences of aging by suggesting these characteristics are universal, natural and inevitable. These assumptions are powerful in creating a knowledge base for health and social welfare professionals who work with older people in particular medical settings such as a hospital or general doctors surgery and also for social workers (Powell 2009).

These new forms of social regulation were also reflected in the family and the community. Hence, modern systems of social regulation have become increasingly blurred and wide-ranging (Powell, 2009). Increasingly, modern society regulated the populational construct by sanctioning the knowledge and practices of the new human sciences—particularly psychology and biology. These are called gerontological "epistemes" which are "the total set of relations that unite at a given period, the discursive practices that give rise to epistemological figures, sciences and possibly formalized systems" (Foucault, 1972: 191). The "psy" complex or bio-medical epistemes refers to the network of ideas about the "nature" of individuals, their perfectability, the reasons for their behavior and the way they may be classified, selected and controlled. It aims to manage and improve individuals by the manipulation of their qualities and attributes and is dependent upon scientific knowledge and professional interventions and expertise. Human qualities are seen as measurable and calculable and thereby can be changed and improved. The new human sciences had as their central aim the prediction of future behavior (Powell, 2009).

Powell and Biggs (2000) suggest that a prevailing ideology of ageism manifests itself in the bio-medical model via its suggestion that persons with such biological traits have entered a spiral of decay, decline and deterioration. Along with this goes certain assumptions about the ways in which people with outward signs of aging are likely to think and behave. For example, there are assumptions that "older people are poor drivers" or that older people have little interest in relationships that involve sexual pleasure that are all explained away by "decline" and "deterioration" master narratives that comprise an aging culture. The effects of the "decline" and "decay" analogies can be most clearly seen in the dominance of medico-technical solutions to the problems that aging and even an "aging population" (Phillipson, 1998) is thought to pose. Here, the bio-medical model has both come to colonize notions of age and reinforce ageist social prejudices to the extent that "decline" has come to stand for the process of aging itself (Powell, 2001).

Estes and Binney (1989 cited in Powell, 2001) have used the expression "biomedicalization of aging" which has two closely related narratives: (1) the social construction of aging as a medical problem, and (2) ageist practices and policies growing out of thinking of aging as a medical problem. They suggest:

> "Equating old age with illness has encouraged society to think about aging as pathological or abnormal. The undesirability of conditions labeled as sickness or illness transfer to those who have these conditions, shaping the attitudes of the persons themselves and those of others towards them. Sick role expectations may result in such behaviors as social withdrawal, reduction in activity, increased dependency and the loss of effectiveness and personal control—all of which may result in the social control of the elderly through medical definition, management and treatment" (Estes and Binney, 1989, quoted in Powell, 2001: 119).

These authors highlight how individual lives and physical and mental capacities that were thought to be determined solely by biological and psychological factors, are, in fact, heavily influenced by social environments in which people live. This remains invisible to the bio-medical approach because they stem from the societal interaction before becoming embedded and recognizable as an "illness" in the aging body of the person. For example, in the “sociology of emotions” the excursion of inquiry has proposed that “stress” is not only rooted in individualistic emotional responses but also regulated, classified, and shaped by social norms of western culture (Powell and Biggs, 2000).

This type of research enables the scope of aging to be broadened beyond biomedical individualistic accounts of the body. On this basis alone, sociology invites us to recognize that aging is not only a socially constructed problem by bio-medical sciences but also the symptomatic deep manifestation of underlying relations of power and inequality that cuts across and through age, class, gender, disability and sexuality (Powell and Biggs, 2000; Powell, 2001).

At this level of analysis, sociology addresses bio-medicine as one of the elements of social control and domination legitimated through power/knowledge of “experts” (Powell and Biggs, 2000). Such expert formation has also been labeled as agist (Bytheway, 1993). Ageism is where the assumptions made about old age are negative, which treats older people not as individuals but as a homogenous group, which can be discriminated against (Bytheway, 1993).

Chinese society uses age categories to divide this ongoing process into stages or segments of life. These life stages are socially constructed rather than inevitable. Aging, too, is a production of social category. At any point of life span, age simultaneously denotes a set of social constructs, defined by the norms specific to a given society at a specific point in history. Thus, a specific period of life: infancy, childhood, adolescence, adulthood, middle age or old age is influenced by the structural entities of a given society. Therefore, aging is not to be considered the mere product of biological-psychological function rather a consequence of sociocultural factors and subsequent life-chances. Indeed, society has a number of culturally and socially defined notions of what Phillipson (1998) calls the “stages of life.” However, a fundamental question is how bio-medical gerontology has stabilized itself with a positivist discourse that not only reflects history but also the total preoccupation with the “problems” of aging that have important implications for older people and health lifestyles in China.

BIO-MEDICINE, FAMILY CARE, AND AGING: IMPLICATIONS FOR CHINA

The dominant bio-medical discourse of aging in China dwells on the processes of physical deterioration associated with becoming older. In this perspective, the aging body has to deal with increased levels of incapacity, both physical and mental, and becomes increasingly dependent on younger others for sustenance and survival; it is the family through informal care that has to provide care of older people who may have illnesses, according to the Law on the Rights and Interests of the Elderly, introduced by the PRC in 1996. The bio-medical problematization of aging has secreted wider questions of power and inequality; especially influential is occidental modernity. A powerful discourse is thus developed which follows that of the West, via notions of "social inclusion" and "family care," and the all-important role of the consumer in buying products for the elderly, from disability aids through to private pensions (Powell and Cook, 2001). The latter suggest that this process constructs the aging body as a site of surveillance by the Chinese state, constructing them as, following Foucault, objects of power and knowledge in which "its your age" is the prevailing authority response to the elderly "customer."

Powell and Cook (2001) have further noted that older people will be increasingly probed for social, psychological and economic factors such as "frailty" or "expected level of supervision." "There are indications, for example, that where care homes are provided, these are for the more active elderly, rather than those in greatest need" (Powell and Cook, 2001: 7).

This Foucauldian point has been borne out via an chapter on Shanghai (China Daily, 2004) with "most nursing homes in Shanghai have entry criteria that target a narrow minority of elderly people. Some admit only those who are capable of independent living while others accept only bedridden patients. While dementia is a common condition among the elderly, those afflicted by it are generally excluded by the criteria." Even more seriously, if a patient's condition changes according to these criteria, he or she is forced to leave the home. "The lack of a continuum of care creates devastating situations for the patients and their families." Further, nursing homes have only minimal level of medical support available, and patients are transferred to hospitals too readily if they have an ailment much beyond the common cold. The patient can then lose their place in the nursing home if their bed is transferred and thus be subject to further stress. "Preventive care, physical therapy, and spiritual care, which are crucial components of care for the elderly, are generally overlooked. Many nursing homes do not provide such services out of concerns for cost or accident liabilities." The Shanghai chapter also notes, damningly, that:

> "The financial burden of long-term care accumulates on an elderly population already enduring tighter budget constraints because of retirement and unemployment. In the absence of government subsidy, the higher fees charged by selfsustained nursing homes deprive elderly people with limited financial means of their access to care" (China Daily, 2004).

In the light of these and other issues the PRC government is attempting to change the ways in which the elderly are perceived, via campaign slogans such as "respect the elderly" and "people first." The former campaign seeks to encourage younger people to visit the elderly on a regular basis in order to reduce the sense of isolation that the elderly can feel, to look out for their needs out on the streets and to generally raise awareness of the situation of older people. There is a resonance here with ancient Confucian tradition in which the elderly were venerated. There have also been attempts to encourage younger people to think about "healthy aging," but this is meant in terms of ensuring that they themselves have adequate financial provision as they age. "People first" is the attempt to recognize that "aging is an individual-specific process" and that "a functional healthcare system for the elderly should integrate all aspects of care … emphasizing and fulfilling individual needs and preferences" (CD, 2004).

Older people in rural areas are more likely to have to face emigration of their children to the cities as China's urbanization proceeds apace. This can leave them physically and socially isolated in a remote rural area, no longer able to rely upon their family to look after them in their old age, as was once the tradition.

Indeed, as social norms and values change, younger people may no longer be willing, even when they are able, to support elderly parents, and in recent years the law has been used to take children to court in order to force them to support their parents. For example, a new law came into force on October 1, 1996 on "The Rights and Interests of the Elderly" that explicitly states that: "the elderly shall be provided for mainly by their families, and their family members shall care for and look after them" (Du and Tu, 2000).

Notwithstanding the legal process, at its most extreme, the concerns of elderly people are expressed via suicide—gerontocide. The elderly can struggle in the face of the massive social changes that China is facing, and the abandonment of the tradition in which they themselves would have looked after their own parents and grandparents. They may feel so stressed and alienated that suicide seems the preferred option.

For example, a 76 year old man blew himself up in a courtroom in protest during a case against his family, who had offered only 350 Yuan a month to support him when 600 Yuan was required (8 Yuan = 1 USD) (Cook and Murray 2001). In a society of rapid transformation, older people in particular may be vulnerable to the sense of abandonment within a more materialistic and selfish new world epitomized by the forces of global capitalism and seeping impingement into day to day living of older people in China.

Declining to Decline—Active Aging?

At the other end of the scale is the active elderly, probably the ideal state for all elderly people. Briefly, older people traditionally were more likely to be active within rural areas of China, in part because they had to be in order to maintain their livelihood. This particular tradition continues today; official data for the 1990s showed that 26 percent of people in rural areas still depended on their own labor earnings compared to only 7 percent in urban areas.

This is not about advocating that elderly people should have to work to continue to earn a living, but we are suggesting that an active lifestyle be promoted where possible. In the cities of China today, it is heartening to see the colonization of open space throughout the hours of daylight and even into the evening by the elderly who are engaged in a wide variety of activities, from the traditional (such as taijiquan and qigong), walking one's pet bird, traditional dance or poetry writing using brushes dipped in water only, illustrating the ephemeral and passing nature of life itself). Hopefully, the increased pollution which China's cities face will not erode the potential gains from these and other activities for older people. An alternative discourse on aging can point to ways in which the elderly can be *deproblematized* as a negative medical, economic, political and social category.

This must begin with appreciation of older people first of all as people rather than as a category (Powell 2006). Older people share, however, apart from their longevity, a wide and deep experience of life itself, and thus of life situations. In China, older people used to be venerated because they were, almost literally, the founts of wisdom, the holders of accumulated knowledge far in advance of the younger members of the family and community. Today, knowledge is far less likely to be oral, and far less dependent on accumulation by the individual. Instead, it is increasingly available at the touch of a button via an Internet search engine, even if there are some restrictions on web provision in the PRC. But, it can be suggested, there should still be a major role for the accumulated wisdom of older people's experiences as carriers of historical wisdom.

In China, there is a growing awareness of the need to have a sophisticated, multidimensional policy to respond to the needs of older people. But there is still a strong bio-medical emphasis on surveillance and control. It could be suggested that policy needs to be driven by the elderly themselves wherever possible. They should be encouraged to define and state their own needs, provided with support when this is required, but the overriding emphasis should be on providing support that fosters active lifestyle and independent living wherever and whenever possible. This means encouragement to the elderly to share their accumulated experience, to provide their oral histories and their views on the momentous changes that they, along with China itself, have lived through. Older people should be valued, and involved in wider society on terms that they themselves desire, recognizing that a wish for privacy and seclusion might be their preference.

Conclusion

Finally, it is worth noting that cases exist where narratives and micro-histories coexist and play a role in producing and strengthening social exclusion. The medical "gaze" refers here to discourses, languages, and ways of seeing that shape the understanding of aging into questions that center on, and increase the power of, the State, and restrict or de-legitimize other possibilities. A consequence is that areas of policy may at first seem tangential to the medical project come to be reflected in its particular distorting mirror. The Chinese "Duomin"—a subcategory of a wider population officially catalogued as fallen people, beggars or ruined households—were seen as inferior and condemned to bear low status on account of a number of beliefs prevailing among mainstream society. The narrative concerns a creational myth that asserts that the Doumin were closely related to Chinese ethnic minorities like the She and Yao and that all these groups shared the belief in Pan-hu, a common dog ancestor. As to micro-histories, we run into different stories which state that the Doumin were either: (1) descendants of Song Dynasty traitors, deserters or prisoners; (2) remnants of antique non-Chinese ethnic groups; (3) foreigners who adopted the customs of Chinese lower social strata; and (4) descendants of domestic slaves. Yet in all cases the Doumins excluded and outcaste status came about as a punishment society bestowed upon them. They were reduced to performing polluting occupations (ox head lanterns making, ironwork, barbers, caretakers, frog-catching, entertaining, among others) and limited to live in segregated quarters outside town. Furthermore, the Doumin were not allowed to study or take public office, nor serve as officers and were obliged to marry among themselves. This suggests that their social identity was a result of the legal status imposed on them and not the other way around. As Hansson (1996: 87) expresses it: "Once fallen people had been labeled as beggars, they had little choice but to conform to the behavior expected from people who had the social identity associated with their legal status." For older people if they are regarded as an inferior category then their behavior will begin to mimic this categorization. "Dependent is as dependent does" is a major danger in the continuation of the dominance of the bio-medical model of aging.

A key point here is that the notion of the "bio-medical gaze," not only draws our attention to the ways that aging has become "medicalized" as a social issue in China, it also highlights the way in which older people are encouraged for as long as possible to "work on themselves" as active subjects. Thus, as Blaikie (1999) has pointed out, older citizens are encouraged to take greater personal responsibility for their health and for extending this period of their active aging. Those who are defined in relation to their health then discover themselves transformed into passive objects of medical power in China. How do we go beyond this in managing old age?

We could suggest macro-social practices have become translated into particular ways of growing old that not only shape what it is to age successfully, but are also adopted by older adults, modified to fit their own life circumstances and then fed back into wider narratives of aging well. Harry R. Moody (1998) coins this as the "Illderly" and "Wellderly" and managing aging experiences is about resistance to dominant discourses of bio-medicine. According to Frank (1998), the personal experience of illness is mediated by bio-medical procedures that shape and contribute to how the older people recognizes their own process of ill health and recover. Katz (2000) notes that the maxim of "activity for its own sake" as a means of managing later life not only reflects wider social values concerning work and non-work, it also provides personal means of control and acts as grounds for resistance. In addition, Phillipson (1998) argues that changes in westernized policies has occurred from seeing old age as a burden to seeing it as an opportunity to promote productive aging. This reflects an attempt to shape acceptable forms of aging whilst encouraging older adults to self-monitor their own success at conforming to the challenging paradigm to hegemony of bio-medicine and its neglect of the agency of older people. What is needed is a re-balance with social paradigms of aging. The next chapter focuses on this.

Chapter 3

THE RELEVANCE OF SOCIAL THEORIES OF AGING

One of the major problems in social gerontology in recent years is that the study of aging is not clearly directed by social theory. Theoretical developments in gerontology have according to Bengston et al. (1997) lagged well behind other social and human science disciplines. George (1995) consolidates this by claiming that gerontological research is "theoretically sterile". In a study of published literature in eight leading social gerontology journals from 1990 to 1995, Bengston et al. (1997) claimed that 80% of the chapters lacked a theoretical framework for their research findings. Bengston et al. (1997) claim that social gerontology has been "itty-bitty" in its development of theoretical frameworks to assist an understanding of an important issue such as 'age': "… it is intellectually irresponsible for a program of research to proceed without … a theory" (Bengston et al. 1997, p.73). This chapter faces up to Bengston et al. (1997) challenge and is concerned with finding theoretical ideas which have informed social understanding of age and aging in recent years. Whilst Bengston et al. (1997) are correct to assert the poverty of theory in social gerontology, it would be folly to suggest that no theories exist. Indeed, there have been some significant theoretical ideas which have influenced understanding of social gerontology in recent years: functionalism, marxism, feminism and postmodernism. Whilst these have increased in terms of scholars understanding of aging, quantitative research findings have dominated the gerontological landscape in publications on aging in the US. It is however important to illuminate the contrasting theories of age and aging, as 80% of research chapters in mainstream social gerontology have not attempted to review or apply these theories in an attempt to understand the philosophical dimensions of adult aging (Bengston et al. 1997). When one assesses the theoretical literature in gerontology as applied to understanding China or global aging, there is a paucity of theory. It would be misleading to suggest no theories exist, so its important to map them out and explore relevance for appreciating human aging.

A Brief History of Theory in Social Gerontology: The Rise of Functionalism

The emergence of the social theories of age and aging can be located to the early post-war years with the governmental concern about the consequences of demographic change and the shortage of younger people in work in USA and UK (Biggs & Powell, 2001). In the post war years, social gerontology emerged as a multidisciplinary field of study which attempted to respond to the social, health and economic policy implications and projections of populational change (Phillipson, 1998). The wide disciplinary subject matter of social gerontology was shaped by significant external forces: first, by state intervention to achieve specific outcomes in health and social policy for older people; secondly, by a socio-political and economic environment which viewed an aging population as creating a 'social problem' for western society in general (Phillipson, 1998). The important point to note is that theories often mirror the norms and values of their creators and their social times, reflecting culturally dominant views of what should be the appropriate way to analyse social phenomena (Turner, 1989). The two functionalist theories which dominated US gerontology in the 1950s and 1960a of Disengagement and Activity theory follow this normative pattern. Both disengagement and activity theories postulate not only how individual behaviour changes with aging, but also imply how it should change.

Functionalist sociology dominated the sociological landscape in the USA from the 1930's up until 1960s (Blaikie 1999). This was seen in the United States via the dominance of the structural-functionalist school via the work of 'disengagement theorists' regarding the study of aging (Phillipson 1998). Such major protagonists of disengagement theory were Cumming and Henry (1961) who looked at how older people should disengage from work roles and prepare for the ultimate disengagement: death (Powell, 2000; 2001). They also proposed that gradual withdrawal of older people from work roles and social relationships is both an 'inevitable' and 'natural process':

> '...withdrawal may be accompanied from the outset by an increased preoccupation with himself: certain institutions may make it easy for him' (Cumming and Henry 1961: 14).

Cumming and Henry (1961) advocate that the process of disengagement is inevitable, rewarding and universal process of mutual withdrawal of the individual and society from each other with advancing age - was normal and to be expected. This theory argued that it was beneficial for both the aging individual and society that such disengagement takes place in order to minimise the social disruption caused at an aging person's eventual death (Neurgarten 1998).

Retirement is an illustration of the disengagement process, enabling older person to be freed of the roles of an occupation and to pursue other roles not necessarily aligned to full-pay of economic generation. Through disengagement, Cumming and Henry argued, society anticipated the loss of aging people through death and brought "new blood" into full participation within the social world (cited in Katz 1996). Bronley (1966: 136) further portends 'in old age, the individual is normally disengaged from the main streams of economic and community activity'. Not surprisingly for Bromley (1966 quoted in Bond and Coleman 1993: 44) 'The (disengagement) process is graded to suit the declining biological and psychological capacities of the individual and the needs of society'. In order to legitimise its generalisations, disengagement theory self-praised itself to objective and value-free rigour of research methods: survey and questionnaire methods of gerontological inquiry. In a sense, by arguing for 'disengagement' from work roles under the guise of objectivity based on scientific predisposition is a very powerful argument for governments to legitimise boundaries of who can work and who cannot based on age (Powell 1999).

The second functionalist theory, Activity theory is a counterpoint to disengagement theory, since it claims a successful 'old age' is can be achieved by maintaining roles and relationships (Powell, 2001). Activity theory actually pre-dates disengagement theory. In the 1950s Havighurst and Albrecht (1953 cited in Katz 1996) insisted aging can be lively and creative experience. For activity theorists, disengagement is not a natural process as advocated by Cumming and Henry. For activity theorists, disengagement theory is inherently agist and does not promote in any shape or form 'positive aging'. Nevertheless, Activity theory neglects issues of power, inequality and conflict between age groups. An apparent 'value consensus' may reflect the interests of powerful and dominant groups within society who find it advantageous to have age power relations organised in such a way. Whilst Phillipson (1998) sees such functionalist schools as important in shaping social theory responses to them, such functionalist theories 'impose' a sense of causality on aging by implying you will either 'disengage' or will be 'active'. Such theories of aging are very macro orientated and fail to resolve tensions within age-group relations which impinge upon the inter-connection of 'race', class and gender with age (Powell, 1999).

Aging, Political Economy and the Politics of Distribution

As an intellectual backdrop against functionalist theoretical dominance, Political Economy of Old Age emerged as a critical theory in both sides of the Atlantic. Political economy drew from Marxian insights in analysing the capitalist complexity of modern society and how old age was socially constructed to foster the needs of the economy (Estes 1979). This critical branch of Marxist gerontology grew as a direct response to the hegemonic dominance of structural functionalism in the form of disengagement theory, the biomedical paradigm and world economic crises of the 1970s. As Phillipson (1998) points out in the UK huge forms of social expenditure were allocated to older people. Consequently, not only were older people viewed in medical terms but in resource terms by governments. This brought a new perception to attitudes to age and aging. As Phillipson (1998: 17) teases out:

> 'Older people came to be viewed as a burden on western economies, with demographic change... seen as creating intolerable pressures on public expenditure'.

Hence, the major focus is an interpretation of the relationship between aging and the economic structure. In the USA, Political Economy theory was pioneered via the work of Estes (1979), and Estes, Swan and Gerard (1982). Similarly, in the UK, the work of Walker (1981), Townsend (1981) and Phillipson (1982) added a critical sociological dimension to understanding age and aging in advanced capitalist societies. For Estes (1979) political economy challenges the ideology of older people as belonging to a homogenous group unaffected by dominant structures in society. Estes (1979) claims political economy focuses upon an analysis of the state in contemporary societal formations. Estes looks to how the state decides and dictates who is allocated resources and who is not. This impinges upon retirement and pensions. As Phillipson (1982) points out, the retirement experience is linked to the timing of economic reduction of wages and enforced withdrawal from work has made many older people in the UK in a financially insecure position. Hence, the state can make and break the 'minds and hearts' of its populace. Phillipson (1982, 1986) considers how capitalism helps socially construct the social marginality of older people in key areas such as welfare delivery. The important argument Phillipson (1998) makes is that inequalities in the distribution of resources should be understood in relation to the distribution of power within society, rather than in terms of individual variation. Similarly, in the USA, Estes, Swann and Gerard (1982) claims that the state is using its power to transfer responsibility for welfare provision from the state and onto individuals.

Sociological Interpretations of Age and Gender

In recent years, there has been an acceleration of Feminist insights into understanding age and gender as identity variables of analysis (Arber & Ginn 1991 and 1995). There are two important issues: first, power imbalances shape theoretical construction; second, a group's place within the social structure influences theoretical attention they are afforded. Henceforth, because older women tend to occupy a position of lower class status, especially in terms of economic status than men of all ages and younger women, they are given less theoretical attention (Arber & Ginn, 1995). According to Acker (1988 cited in Arber and Ginn 1991) in all known societies the relations of distribution and production are influenced by gender and thus take on a gendered meaning. Gender relations of distribution in capitalist society are historically rooted and are transformed as the means of production change. Similarly, age relations are linked to the capitalist mode of production and relations of distribution. "Wages" take on a specific meaning depending on age. For example, teenagers work for less money than adults, who in turn work for less money than middle-aged adults. Further, young children rely on personal relations with family figures such as parents. Many older people rely on resources distributed by the state.

Older women are viewed as unworthy of respect or consideration (Arber and Ginn 1991). Itzin claims the double standard of aging as arising from the sets of conventional expectations as to age-pertinent attitudes and roles for each sex which apply in patriarchal society. These are defined by Itzin as a male and a female 'chronology', socially defined and sanctioned so that the experience of prescribed roles is sanctioned by disapproval. Male chronology hinges on employment, but a woman's age status is defined in terms of events in the reproductive cycle.

Arguably, Arber & Ginn (1991) claim because women's value is exercised the awareness of a loss of a youthful appearance brings social devaluation; vulnerability to pressure is penetrated by cosmeticisation. Daly (cited in Arber and Ginn 1991) draws a mirror image between western cosmetic surgery and the genital mutilation carried out in some African societies: both cultured practices demonstrate the pressure on women to comply with male standards of desirability and the extent of male domination. For older black women, the ideal of 'beauty' portrayed by white male culture was doubly distant and alienating, until growing black consciousness subverted disparaging language and argued 'black is beautiful'.

It is perhaps emblematic of contemporary western society that aging marginalises the experiences of women through an inter-connected oppression of gender and aging. The reason for this as Arber and Ginn (1991) claim, is that patriarchal society exercises power through the chronologies of employment and reproduction, and through the sexualised promotion of a 'youthful' appearance in women.

AGING IN A POSTMODERN WORLD?

In the 1990s, there has been a vast interest in Postmodern perspectives of age and aging identity underpinned by discourses of "better lifestyles" and increased leisure opportunities for older people due to healthier lifestyles and increased use of bio-technologies to facilitate the longevity of human experiences (Blaikie 1999; Featherstone & Hepworth 1993, Featherstone & Wernick 1995 and Powell & Biggs 2002). The intellectual roots of 'postmodern gerontology' derive from Jaber F. Gubrium's (1975) superb analysis of the discovery of Alzheimer's disease in the USA and the establishment of boundaries between 'normal' and pathological aging; old age is seen as a "mask" which conceals the essential identity of the person beneath. The view of the aging process as a mask/disguise concealing the essentially youthful self beneath is one which appears to be a popular argument (Featherstone & Hepworth 1989, 1993). When asked at the age of 79 to describe what it felt like to be old, the author J.B. Priestley replied:

> 'It is as though, walking down Shaftesbury Avenue as a fairly young man, I was suddenly kidnapped, rushed into a theatre and made to don the grey hair, the wrinkles and the other attributes of age, then wheeled on stage. Behind the appearance of age I am the same person, with the same thoughts, as when I was younger' (Puner 1978: 7).

There are two underlying issues for Featherstone and Hepworth (1993) which should be understood as the basis for understanding postmodern gerontology. Firstly, the the mask alerts social gerontologists to the possibility that a tension exists between the external appearance of the body and face and functional capacities and the internal or subjective sense of experience of personal identity which is likely to become prominent as aging travels through the lifecourse.

Secondly, older people are usually 'fixed' to roles without resources which does not do justice to the richness of their individual experiences and multi-facets of their personalities. Idealistically, Featherstone and Hepworth argue that a postmodern perspective would deconstruct such realities and age should be viewed as fluid with possibilities not constrained by medical model decline discourses.

According to Powell and Biggs (2000) the use of new technologies to modify the appearance of aging identity exemplifies postmodernity. To paraphrase Morris (1998) technologies here hold out the promise of 'utopian bodies' or perfect bodies. Indeed, Haraway's (1991) (cited in Powell and Biggs 2002) reference to cyborgic fusion of biological and machine entities has been taken up by postmodern gerontology. The list of technologies available extends beyond traditional conceptualizations of the body to include virtual identities created by and reflected in the growing number of 'silver surfers' using the Internet as a free-floating form of identity management (Powell & Biggs, 2002). Thus Featherstone and Wernick (1995: 3) claim that it is possible to' Re-code the body itself 'as biomedical and information technologies make available' the capacity to alter not just the meaning, but the very material infrastructure of the body. Bodies can be re-shaped, remade, fused with machines, empowered through technological devices and extensions'.

The use of diet and exercise as techniques specifically related to later adulthood, is closely related to the growth of leisure and a lifestyle approach to the creation of late life identities (Powell & Biggs 2000 and 2002). It therefore resonates beyond the simple fuelling and repair of the bodily machine to include a continual re-creation of the self within a particular social discourse.

CONCLUSION

These theories have been at the forefront of understanding old age in US, UK and Australasian academies. Taken together, these theoretical currents have been influential in providing social gerontology with a rich social dimension. Such social theories have been used also to analyse pressing social issues such as, elder abuse, the gendered nature of age, the politics of power relations between older people and state/society and community care. The purpose of this chapter has been to amalgamate the key ideas of social theories of age in order to stress the importance of social philosophy to understanding age and aging. The task for social gerontology is to integrate it with different cultures and to examine new global questions of aging.

Chapter 4

THE "SILVER TSUNAMI": FROM CHINA TO GLOBAL AGING

Demographers have forecasted quite remarkable increases in the general population of the globe, which will also be reflected in increasing aging populations. This chapter explores how globalization and its structural economic and social forces throws into flux the policies and practices of individual nation states to address social, economic and political issues for older people focusing on pensions and health and social care. The chapter will examine specific areas of the globe such as America, Europe, Asia and Africa before we discuss some of the key challenges and consequences of global aging for global society. It highlights how research needs to move from being state centred to one of which acknowledges global forces and the impact of populational aging.

INTRODUCTION

The aim of this chapter is to analyse the expansion in the proportion of older people across the globe, and to highlight the main social and economic forces causing this.

Indeed, there is little doubt that the rapid increase in population aging across the globe is signalling some striking demographic changes (Gruber and Wise 2004). Amongst policy makers, there is concern about population aging and its consequences for nation states, for sovereign governments and for individuals (Estes, Biggs and Phillipson 2003). The United Nations estimates that by the year 2025, the global population of those over 60 years will double, from 542 million in 1995 to around 1.2 billion people (Krug, 2002:125). The global population age 65 or older was estimated at 461 million in 2004, an increase of 10.3 million just since 2003. Projections suggest that the annual net gain will continue to exceed 10 million over the next decade—more than 850,000 each month. In 1990, 26 nations had older populations of at least 2 million, and by 2000, older populations in 31 countries had reached the 2 million mark (Chen and Powell, 2011). UN projections to 2030 indicate that more than 60 countries will have at least 2 million people age 65 or older.

While today's proportions of older people typically are highest in more developed countries, the most rapid increases in older populations are actually occurring in the less developed world (Krug 2002). Between 2006 and 2030, the increasing number of older people in less developed countries is projected to escalate by 140% as compared to an increase of 51% in more developed countries (Powell 2011). A key feature of population aging is the progressive aging of the older population itself. Over time, more older people survive to even more advanced ages. The forecast rise in the number of older people aged 75+ over the next 20 years will lead to an expansion of demand for health, housing accommodation and pensions for aging populations and is thus of crucial importance for governments, policy makers, planners, and researchers in all nation states. On a global level, the 85-and-over population is projected to increase 151% between 2005 and 2030, compared to a 104% increase for the population age 65 and over and a 21% increase for the population under age 65 (Bengston and Lowenstein 2004). The most striking increase will occur in Japan: by 2030, nearly 24% of all older Japanese are expected to be at least 85 years old. As life expectancy increases and people aged 85 and over increase in number, four-generation families may become more common (Powell 2011).

The age structure of the population has changed from one in which younger people predominated to a global society in which people in later life constituted a substantial proportion of the total population (Powell, 2005). Transformations in the age profile of a population are a response to political and economic structures. Older people in particular constitute a large section of populations in western society in particular but the percentage of pensionable age is projected to remain at 18% until 2011 when it becomes 20% and rising to 24% in 2025 (Chen and Powell 2011).

At the same time, there is a stigmatisation of such increasing populational numbers by agist stereotypes. In relation to public services that have to be paid for by 'younger' working people, the percentage of the population has been used to signify such 'burdensome' numbers by the State (Estes, Biggs and Phillipson 2003). Dependency rates, that is the number of dependants related to those of working age, have in fact altered little over the past 100 years. The reason there has been so little change during a period of so called rapid aging populations is that there has been a fall in the total fertility rate (the average number of children that would be born to each woman if the current age-specific birth rates persisted throughout her child-bearing life).

In advanced capitalist or First World countries, declines in fertility that began in the early 1900s have resulted in current fertility levels below the population replacement rate of two live births per woman. Perhaps the most surprising demographic development of the past 20 years has been the pace of fertility decline in many less developed countries (Giddens 1993). In 2006, for example, the total fertility rate was at or below the replacement rate in 44 less developed countries (Cook and Powell, 2007). Most of the more developed nations have had decades to adjust to this change in age structure. For example, it took more than a century for France's population age 65 and over to increase from 7% to 14% of the total population. In contrast, many less developed or Third World countries are experiencing rapid increases in the number and percentage of older people, often within a single generation. The same demographic aging process that unfolded over more than a century in France will occur in two decades in Brazil (OECD 2007). In response to this compression of aging, institutions must adapt quickly to accommodate a new age structure. Some less developed nations will be forced to confront issues, such as social support and the allocation of resources across generations, without the accompanying economic growth that characterized the experience of aging societies in the West. In other words, some countries 'may grow old before they grow rich' (Cook and Powell 2010).

Globalisation has also produced a distinctive stage in the social history of populational aging, with a growing tension between nation state-based solutions (and anxieties) about growing old and those formulated by global institutions (Powell 2011). Globalisation, defined here as the process whereby nation-states are influenced (and sometimes undermined) by trans-national actors (Powell 2010), has become an influential force in shaping responses to population aging. Growing old has, itself, become relocated within a trans-national context, with international organisations (such as the World Bank and International Monetary Fund) and cross-border migrations, creating new conditions and environments for older people.

Aging can no longer just be viewed as a 'national' problem but one that affects trans-national agencies and communities. Local or national interpretations of aging had some meaning in a world where states were in control of their own destiny (Estes, Biggs and Phillipson, 2003). They also carried force where social policies were being designed with the aim or aspiration of levelling inequalities, and where citizenship was still largely a national affair (and where there was some degree of confidence over what constituted 'national borders') (Chen and Powell 2011). The crisis affecting each of these areas, largely set in motion by different aspects of globalisation, is now posing acute challenges for understanding ‘global aging’ in the twenty-first century.

If these examples illustrate the complexity and impact of global aging – then it may be pertinent to highlight how populational aging is impacting more specifically across different continents across the globe. The following section looks at aging in four key areas across the globe: (i) Americas (ii) Asia (iii) Europe and (iv) Africa. These areas illustrate how population growth is impacting and creating social implications concerning health and disease as well as economic concerns relating to the labour market and pensions.

AGING POPULATIONS IN THE AMERICAS

Since the turn of the last century, the life expectancy of people born in North America has increased by approximately 25 years and the proportion of persons 65 years or older has increased from 4% to over 13% (Estes and Associates 2001). By the year 2030, one in five individuals in the U.S. is expected to be 65 years or older and people age 85 and older make up the fastest growing segment of the population. In 2000, there were 34 million people aged 65 or older in the United States that represented 13% of the overall population (Estes and Associates 2001). By 2030 there will be 70 million over 65 in the United States, more than twice their number in 2000. 31 million people, or 12 percent of the total population, are aged 65 and older. In another 35 years, the elderly population should double again (Cook and Powell 2010). The aging population is not only growing rapidly, but it is also getting older:

> ‘In 1990, fewer than one in ten elderly persons was age 85 or older. By 2045, the oldest old will be one in five. Increasing longevity and the steady movement of baby boomers into the oldest age group will drive this trend’ (Longino, 1994: 856).

The percentage of oldest old will vary considerably from country to country. In the United States, for example, the oldest old accounted for 14% of all older people in 2005. By 2030, this percentage is unlikely to change because the aging baby boom generation will continue to enter the ranks of the 65-and-over population (Bengston and Lowenstein 2004). This is obviously causing much concern among policymakers but Longino (1994), for instance, believes that thanks to better health, changing living arrangements and improved assistive devices, the future may not be as negative as we think when we consider an aging population.

It will be different, however, not least because people currently divorced constitute a small proportion of older populations. This will soon change in many countries as younger populations with higher rates of divorce and separation, age. In the United States, for example, 9% of the 65-and-over population is divorced or separated compared to 17% of people age 55 to 64 and 18 percent of people age 45 to 54 (Manton and Gu 2001). This trend has gender-specific implications: in all probability nonmarried women are less likely than nonmarried men to have accumulated assets and pension wealth for use in older age, while older men are less likely to form and maintain supportive social networks.

Shoring up public pensions is hardly the only avenue nations in North and South America are exploring. In many countries, privately managed savings accounts have been strongly advocated (Estes and Associates 2001). Two decades ago, nearly every South American nation had pay-as-you-go systems similar to the US Social Security system. Some granted civil servants retiring in their 50s full salaries for life. Widening budget deficits changed that. In 1981, Chile replaced its public system with retirement accounts funded by worker contributions and managed by private firms. The World Bank encouraged 11 other Latin nations to introduce similar features. For example, in Chile the government addressed its fiscal budget deficit by mobilizing a $49 billion of pension-fund assets that make it easier for companies and corporations to fund investments in the local currency with bond offerings, and most workers have some retirement benefits from this (OECD 2007). At the same time, the downside has been those people who cannot afford a private pension have been left to a low state pension which has intensified poverty (Estes and Associates 2001); an enduring feature of all nation states in America. For the future, there is no safety guarantee that private pension schemes are protected and pay out for people who invest their savings in such provision. In a de-regulated US pension system, the issue of corporate crime has highlighted the continuing problem of private pension provision. In one example, this was seen clearly with the energy corporation of Enron's embezzlement of billions of dollars of employees private pension schemes (Powell, 2005). This debate amounts to a significant global discourse about pension provision and retirement ages, but one which has largely excluded perspectives which might suggest an enlarged role for the state, and those which might question the stability and cost effectiveness of private schemes. The International Labour Organisation (ILO) concluded that investing in financial markets is an uncertain and volatile business: that under present pension plans people may save up to 30 per cent more than they need, which would reduce their spending during their working life; or they may save 30 per cent too little - which would severely cut their spending in retirement (Phillipson, 1998; Estes, Biggs and Phillipson, 2003).

Holtzman (1997), in a chapter outlining a World Bank perspective on pension reform, has argued for reducing state pay-as-you-go (PAYG) schemes to a minimal role of basic pension provision. This position has influenced both national governments and transnational bodies, such as the International Labour Organisation (ILO), with the latter now conceding to the World Bank's position with their advocacy of a mean-tested first pension, the promotion of an extended role for individualized and capitalized private pensions, and the call for Organisation for Economic Co- operation and Development (OECD) member countries to raise the age of retirement.

There is also the impact of (Intergovernmental Organizations) IGOs on the pensions debate in South America. The function of such arguments is to create a climate of fear, of inevitability and scientific certainty that public pension provision will fail. In so far as this strategy succeeds it creates a self-fulfilling prophecy. If people believe the 'experts' who say publicly sponsored PAYG systems cannot be sustained, they are more likely to act in ways that mean they are unsustainable in practice. Certainly, in Europe and elsewhere, the state pension is an extremely popular institution. To have it removed or curtailed creates massive opposition. Only by demoralising the population with the belief that it is demographically unsustainable has room for the private financiers been created and a mass pensions market formed.

Increasingly, the social infrastructure of welfare states is being targeted as a major area of opportunity for global investors. The World Bank has expressed the belief that the public sector is less efficient in managing new infrastructure activities and that the time has come for private actors to provide what were once assumed to be public services. This view has been strongly endorsed by a variety of multinational companies, especially in their work with the World Trade Organisation (WTO). The WTO enforces more than twenty separate international agreements, using international trade tribunals that adjudicate disputes. Such agreements include the General Agreement on Trade in Services (GATS), the first multilateral legally enforceable agreement covering banking, insurance, financial services and related areas (Estes, Biggs and Phillipson, 2003).

AGING IN ASIA

Asia has the fastest increase in the aging population in the world. Du and Tu (2000) suggest that China in particular has been identified as having four 'unique characteristics' of populational aging: firstly, the issue of 'unprecedented speed' with the proportion of aging population is growing faster than Japan, the country previously recognized as having the fastest rate, and much faster than nations in Western Europe; secondly, the 'early arrival of an aging population' before modernization has fully taken place has welfare implications. 'It is certain that China will face a severely aged population before it has sufficient time and resources to establish an adequate social security and service system for the elderly' (Du and Tu, 2000, 79); thirdly, Du and Tu suggest 'fluctuations in the total dependency ratio'. They claim the Chinese government estimates are that the country will reach a higher 'dependent burden' earlier in the twenty-first century than was previously forecast; and fourthly, Du and Tu refer to the 'strong influence of the government's fertility policy and its implementation on the aging process'. Here they explicitly refer to the Single Child Family Policy (SCFP) in China which means fewer children being born, but with more elderly people a conflict arises between the objectives to limit population increase and yet maintain a balanced age structure.

The combination of such factors means that the increased aging population is giving rise to serious concerns among Chinese policy makers.

Kim and Lee (2007) claim the growing elderly population is beginning to exert pressure on the East Asian countries economies. Three decades ago, major industrialized countries have begun to grapple with the similar problem. With increasing drop in fertility rates, more East Asian economies such as Japan, Hong Kong, South Korea, Singapore and Taiwan are expected to turn into "super-aging societies" by 2025 (Kim and Lee 2007). However, the magnitude of the future impact depends on the (in)ability of individual economies to resolve the demographic changes problem through increased privatisation, pension reforms, a migration on more productive countries and extension of retirement age. Like western countries, Asia will ultimately have to tackle issues related to pension reform and the provision of long term health care services (Cook and Powell, 2007).

For Japan, the basic statistical reality of its demographic profile is escalating. Already, 17 of every 100 of its people are over 65, and this ratio will near 30 in 15 years. From 2005 to 2012, Japan's workforce is projected to shrink by around 1% each year - a pace that will accelerate after that. Economists fear that, besides blowing an even bigger hole in Japan's underfunded pension system (Cook and Powell, 2007), the decline of workers and young families will make it harder for Japan to generate new wealth.

The future challenge of providing for the elderly is especially urgent in the world's two biggest nations - India and China. Only 11% of Indians have pensions, and they tend to be civil servants and the affluent. With a young population and relatively big families, many of the elderly population still count on their children for support. This is not the case in China. By 2030, there will be only two working-age people to support every retiree. Yet only 20% of workers have government- or company-funded pensions or medical coverage (Chen and Powell 2011). However, as a counterbalance to such a gloomy perspective, 'Chindia' (China and India taken together) is currently accumulating vast wealth as a result of global change, wealth that could potentially be redirected for the support of their elderly populations.

THE POPULATIONAL STRUCTURE OF EUROPE

The population structure of Western European countries has changed since the turn of the 20th century. Whereas in 1901, just over 6% of the population were at or over current pension age (65 in the UK for men and women), this figure rose steadily to reach 18% in 2001 (Powell, 2011). At the same time, the population of younger people under age 16 fell from 35% to 20%. As European countries reach a relatively high level of population aging, the proportion of workers tends to decline. European countries, including France, Germany, Greece, Italy, Russia, and the Ukraine, already have seen an absolute decline in the size of their workforce. And in countries where tax increases are needed to pay for transfers to growing older populations, the tax burden may discourage future workforce participation. The impact on a Nation States gross domestic product will depend on increases in labor productivity and that State's ability to substitute capital for labor. Less developed countries can shift their economies from labor-intensive to capital-intensive sectors as population aging advances. Options for more European nation states may be more constrained. The 'rolling back' of pensions promises is just one symptom of a shift in European history: the 'graying of the baby-boom generation' (Phillipson 1998). The percentage of 60-year-olds and older are growing 1.9% a year. This is 60% faster than the overall global population. In 1950 there were 12 people aged 15 to 64 to support each one of retirement age. Currently, the global average is nine. It will be only four-to-one by 2050 (Powell, 2005). By then numbers of older people will outnumber children for the first time. Some economists fear this will lead to bankrupt pensions and lower living standards. It is interesting that in Germany this fear is becoming a battleground for political electioneering. For example, Germany has the highest population in Europe and the third oldest population in the world, which presents both critical questions on public finances to provide pensions and healthcare and an opportunity for innovations in the marketplace. Currently, aging has started to figure prominently in political discussions amongst political parties as they vie for the elderly vote. The current Merkel administration (2007-) has been criticized for increasing pensions while opponents talk about a "war of generations" requiring young people to pay for taxation for elder care (Powell 2011).

The trend has drawn further attention across Europe, where the working-age population will decline by 0.6% this in 2010. By 2025 the number of people aged 15 to 64 is projected to dwindle by 10.4% in Spain, 10.7% in Germany and 14.8% in Italy. But aging is just as dramatic in such emerging markets as China - which is expected to have 265 million 65-year-olds by 2020 - and Russia and Ukraine (Cook and Powell, 2007).

Using evidence from the UK, the percentage of people of working age, that is 16-64, will drop from 64% in 1994 to 58% in 2031 (Powell, 2005). As the number of workers per pensioner decreases there will be pressure on pension provision. This is evident now, in such areas of pensions and long term care, the retreat of the state made evident in the erosion of State Earnings Related Pay are forcing people to devise their own strategies for economic survival in old age (Phillipson 1998). Private pensions are slowly being introduced in order to prevent the 'burden' of an aging population. These are ways in which the State continues to rely on apocalyptic projections such as 'demographic time bomb' about aging populations in order to justify cuts in public expenditure (Powell 2005). Hence, the population of the UK, like that of other European countries, is aging rapidly. There are only enough young people to fill one in three of the new and replacement jobs that will need to be taken up over the next decade. Older people take much of the responsibility for our social and civic life and for the care of children, the sick and the very old in the community. Yet the gap between wealth and poverty, choice and the absence of choice for older people is stark and growing wider (Phillipson 1998). The UK government is at the time of writing seeking to promote a debate over what they envisage as a multi-billion pound deficit that will be found in care for the elderly in future.

POPULATIONAL AGING IN AFRICA

Economic security, health and disability, and living conditions in old age are policy concerns throughout the world, but the nature of the problem differs considerably from continent to continent and between and within countries – especially within Africa.

In Africa older people make up a relatively small fraction of the total population, and traditionally their main source of support has been the household and family, supplemented in many cases by other informal mechanisms, such as kinship networks and mutual aid societies. In 2005, Nigeria ranked among the top 30 countries in the world on the basis of the size of its population age 60 and over. Nigeria had the largest older population in sub-Saharan Africa, with over 6 million people age 60 and over; South Africa had just over 3.4 million. Congo and South Africa are projected to have nearly 5 million older people in 2030. Burkina Faso, Cameroon, Cote d'Ivoire, Madagascar, Mozambique, Niger, Senegal, and Uganda are all projected to have their older populations grow to over one million people by 2030 (Building Blocks 2004). Very little careful empirical research has been undertaken on long-term trends in the welfare of older people, but there are a number of reasons to believe that traditional caring and social support mechanisms in Africa are under increasing strain (OECD 2007).

Located on the least developed and poorest continent, African economies are still heavily dependent on subsistence agriculture, and average income per capita is now lower than it was at the end of the 1960s. Consequently, the region contains a growing share of the world's poor. In addition, reductions in fertility and child mortality have meant that, despite the huge impact of the HIV/AIDS epidemic across much of the region, both the absolute size and the proportion of the population age 60 and over have grown and will continue to grow over the next 30 years (Estes, Biggs and Phillipson 2003).

In Africa, as in other traditional societies such as those in India or China, older people have traditionally been viewed in a positive light, as repositories of information and wisdom. And while African families are generally still intact, development and modernization are closely connected with social and economic changes that can weaken traditional social values and networks that provide care and support in later life. Africa has long carried a high burden of disease, including from malaria and tuberculosis; today it is home to more than 60% of all people living with HIV—some 25.8 million in 2005. The vast majority of those affected are still in their prime wage-earning years, at an age when, normally, they would be expected to be the main wage earners and principal sources of financial and material support for older people and children in their families. Many older people have had to deal with the loss of their own support while absorbing the additional responsibilities of caring for their orphaned grandchildren. Increasingly, then, it appears that African societies are being asked to cope with population aging with neither a comprehensive formal social security system nor a well-functioning traditional care system in place (Building Blocks 2004).

The big issue is that majority of the world's population of older people (61 per cent, or 355 million) live in poorer African countries. This proportion will increase to nearly 70 percent by 2025. For many countries, however, population aging has been accompanied by reductions in per capita income and declining living standards. Epstein (2001) notes that between 1950 and the late 1970s, life expectancy increased by least 10 per cent in every developing country in the world, or on average by about 15 years. However, at the beginning of the twenty- first century, life expectancy remains below fifty in more than ten developing countries, and since 1970 has actually fallen, or has barely risen in a number of African countries (Phillipson 1998). The AIDS epidemic is certainly a major factor here, but development loans requiring the privatization of health care have also had an impact. Epstein (2001) reports, for example, that by the mid- 1990s the African continent was transferring four times more in debt repayment than it spent on health or education. More generally, Help Age International (2000: 8) argue that:

> 'Older people's poverty is still not a core concern in the social, economic and ethical debates of our time. Their right to development is routinely denied, with aging seen as a minority interest or case for special pleading. Poverty and social exclusion remain the main stumbling blocks to the realisation of the human rights of older people worldwide.'

The Social and Economic Challenges of Global Aging

Population aging strains social insurance and pension systems and challenges existing models of social support. It affects economic growth, trade, migration, disease patterns and prevalence, and fundamental assumptions about growing older. Older people's living arrangements reflect their need for family, community, or institutional support. Living arrangements also indicate sociocultural preferences—for example, some choose to live in nuclear households while others prefer extended families (Estes, Biggs and Phillipson, 2003). The number, and often the percentage, of older people living alone is rising in most countries. In some European countries, more than 40% of women age 65 and older live alone (Walker and Naeghele 2000). Even in societies with strong traditions of older parents living with children, such as in Japan, traditional living arrangements are becoming less common. In the past, living alone in older age often was equated with social isolation or family abandonment. However, research in many cultural settings illustrates that older people, even those living alone, prefer to be in their own homes and local communities (Gilleard and Higgs, 2001). This preference is reinforced by greater longevity, expanded social benefits, increased home ownership, elder-friendly housing, and an emphasis in many nations on community care.

Global aging will have dramatic effects on local, regional, and global economies. Most significantly, financial expenditures, labor supply, and total savings will be affected. Changes in the age structures of societies also affect total levels of labor force participation in society, because the likelihood that an individual will be in the labor force varies systematically by age. Concurrently, global population aging is projected to lead to lower proportions of the population in the labor force in highly industrialized nations, threatening both productivity and the ability to support an aging population (Krug, 2002).

Coupled with rapid growth in the young adult population in Third World countries, the World Bank (1994) foresee growing ‘threats’ to international stability pitting different demographic-economic regions against one another. The United Nations (2002) views the relationship between aging populations and labor force participation with panic, recognizing important policy challenges, including the need to reverse recent trends toward decreasing labor force participation of workers in late middle and old age despite mandatory retirement in Western countries such as the UK (Powell, 2005). Social welfare provisions and private-sector pension policies influencing retirement income have a major impact on retirement timing. Hence, a major concern for organizations such as United Nations and World Bank centers on the number of such 'dependent' older people in all developing societies.

Some have argued that the rise of globalisation exerts unequal and highly stratified effects on the lives of older people (Estes and Associates 2001). In the developed world, the magnitude and absolute size of expenditure on programmes for older people has made these the first to be targeted with financial cuts. In Third World countries, older people (women especially) have been amongst those most affected by the privatization of health care, and the burden of debt repayments to the World Bank and the IMF (Estes and Associates 2001). Additionally, globalization as a process that stimulates population movement and migration may also produce changes that disrupt the lives of older people. And one must not forget either that they may comprise up to one-third of refugees in conflict and emergency situations - a figure which was estimated at over 53 million older people worldwide in 2000 (Estes and Associates 2001).

Nation states with extensive social programs targeted to the older population — principally health care and income support programs — find the costs of these programs escalating as the number of eligible recipients grows and the duration of eligibility lengthens due to global pressures (Bengston and Lowenstein 2003). Further, few countries have fully funded programs; most countries fund these programs on a pay-as-you-go basis or finance them using general revenue streams. Governments may be limited in how much they can reshape social insurance programs by raising the age of eligibility, increasing contribution rates, and reducing benefits. Consequently, shortfalls may need to be financed using general revenues.

Different countries age groups have different levels of pace of growth. It is possible for the elements of production—labor and capital—to flow across national boundaries and mitigate the impact of population aging. Studies predict that, in the near term, surplus capital will flow from Europe and North America to emerging markets in Asia and Latin America, where the population is younger and supplies of capital relatively low. In another 20 years, when the baby boom generation in the West has mostly retired, capital likely will flow in the opposite direction (May and Powell, 2007). Traditionally, labor is viewed as less mobile than capital, although migration could offset partially the effects of population aging. Currently, 22 percent of physicians and 12 percent of nurses in the United States are foreign born, representing primarily African countries, the Caribbean, and Southeast Asia

(OECD 2007). The foreign-born workforce also is growing in most OECD countries. Over the next 10 years, the European experience will be particularly instructive in terms of the interplay of aging and migration (OECD 2007). Some pressure groups are now suggesting, for instance, that a rich city like London, that benefits from Ghanaian nurses in the National Health Service, has an ethical obligation to Ghana itself, to provide funds to support that country's health training system because the donot country is losing key personnel.

The life-cycle theory of consumption is that households accumulate wealth during working years to maintain consumption in retirement (Gilleard and Higgs 2001). The total of a country's individual life-cycle savings profiles determines whether households in that country are net savers or nonsavers at any point in time. A country with a high proportion of workers will tend to be dominated by savers, placing downward pressure on the rate of return to capital in that economy. Nation states with older populations will be tapping their savings and driving rates of return higher because of the scarcity of capital (Gilleard and Higgs 2001).

Retirement resources typically include public and private pensions, financial assets, and property. The relative importance of these resources varies across countries. For example, a groundbreaking study revealed that only 3% of Spanish households with at least one member age 50 or older own stocks (shares), compared to 38 percent of Swedish households (Walker and Naeghele 2000). The largest component of household wealth in many countries is housing value. This value could fall if large numbers of older homeowners try to sell houses to smaller numbers of younger buyers. How successfully this transition is managed around the world could determine the rise and fall of nations and reshape the global economy in the era of the post-credit crunch. Two key vehicles of growth are increases in the labor force and productivity. If nation states cannot maintain the size of their labor forces by persuading older workers to retire later then the challenge will be to maintain growth levels. That will be a particular challenge in Europe, where productivity growth has averaged just 1.3% since 1995. By 2024, growth in household financial wealth in the U.S., Europe, and Japan will slow from a combined 4.5% annual reduction now to 1.3%. That will translate into $31 trillion less wealth than if the average age were to remain the same (Cook and Powell, 2007).

Most of Europe's state-funded pension systems encourage early retirement. Now, 85.5% of adults in France retire from employment by age 60, and only 1.3% engage in employment beyond aged 65. In Italy, 62% of adults retire from full-time work by the age of 55. That compares with 47% of people who earn wages or salaries until they are 65 in the U.S. and 55% in Japan (Estes and Associates 2001).

Why the sudden attention to a demographic trend of global aging? In part, it is because the future is already dawning that global trends impact on state power. In South Korea and Japan, which have strong cultural aversions to immigration, small factories, construction companies, and health clinics are relying more on 'temporary' workers from the Philippines, Bangladesh, and Vietnam (OECD 2007). In China, state industries are struggling over how to lay off unneeded middle-age workers when there is no social safety net to support them.

What really has pushed aging to the top of the global agenda, though, are increasing fiscal gaps in part, due to the "global credit crunch" in the U.S., Europe, Japan, and elsewhere that could worsen as populations reach retirement age. While U.S. Social Security is projected to remain solvent until at least 2042, the picture is more acute in Europe. Unlike the U.S., where most citizens also have private savings plans, in much of Europe up to 90% of workers rely almost entirely on public pensions (Walker and Naeghele 2000). Austria guarantees 93% of pay at retirement, for example, and Spain offers 94.7%. Pensions and elder-care costs will increase from 14% of capitalist nations' gross domestic product to 18% by 2050 (Walker and Naeghele 2000).

As people live longer and have fewer children, family structures are also transformed (Bengston and Lowenstein 2004). This has important implications in terms of providing care to older people. Most older people today have children, and many have grandchildren and siblings. However, in countries with very low birth rates, future generations will have few if any siblings. As a result of this trend and the global trend toward having fewer children, people will have less familial care and support as they age (Bengston and Lowenstein 2004). Unless there is a fundamental shift in the views of 'Fortress Europe', Japan and other countries towards immigrants, and an overcoming of entrenched racial or racist attitudes towards migrants, some parts of the globe will be 'elderly heavy' while others will be 'elderly light'. Were migrants made more welcome in richer societies then one could envisage a space of carer flows, with more interactions and movements in either direction to the 'heavy' or 'light' end. Or, for example, one could have elderly relocation in the same way as Japanese elderly are relocating into Thailand into new forms of 'transnational households', in order to seek cheaper care systems for their retirement (Toyota, 2006).

As a consequence of the global dynamics of aging, the changing societies of the post millennium are being confronted with quite profound issues relating to illness and health care, access to housing and economic resources including the retirement experience and pension provision. If demographic trends continue to escalate by 2050 the number of older people globally will exceed the number of young for the first time since formal records began raising questions of the power of the nation state in the context of global aging, and raising further global questions of distribution of power and scarcity of resources to an (global) aging population.

References

Arber, S. and Ginn, J. (Eds.) (1991) *Connecting Gender and Aging: a sociological approach*, Milton Keynes: OUP.

Baltes, M. & Carlestensen, L. (1996). The process of successful aging. *Aging & Society*, 16, 397-422.

Baltes, P. B. (1999). The incomplete architecture of human ontogenesis. *Gerontologie unt Geriatrie*. 32. 6: 433-448.

Bengston, V., Burgess, E, & Parrot, T. (1997). 'Theory, Explanation and a Third Generation of Theoretical Development in Social Gerontology', *Journal of Gerontology: Social Sciences*, 52 (B), 72-88.

Bengston, V.L. and Lowenstein, A. (Eds.) (2004) *Global Aging and Challenges to Families*. New York: De Gruyter

Bengtsen, V. & Schaie, R. (1999). *Handbook of Theories in Gerontology*. New York: Springer.

Biggs, S and Powell, J.L. (1999). 'Surveillance and Elder Abuse: The Rationalities and Technologies of Community Care', *Journal of Contemporary Health*, 4, (1), 43-49.

Biggs, S. (1993). *Understanding Aging*. Milton Keynes: OUP.

Biggs, S. (1996). A family concern: elder abuse in British social policy. *Critical Social Policy*. 16.2:63-88.

Biggs, S. (1999). *The Mature Imagination*. Milton Keynes: OUP.

Blaikie, A. (1999). *Aging and Popular Culture'*. Cambridge: OUP.

Bowl, R. (1986). 'Social work with old people' in Phillipson, C. and Walker, A. (Eds.) *Aging and Social Policy*. London: Gower.

Building blocks (2004) *Africa-wide briefing notes – supporting older carers*, HIV AIDS Alliance and HelpAge International.

Bury, M. (1995). 'Aging, Gender and Sociological Theory' in Arber, S. and Ginn, J. (Eds.) *Connecting Gender and Aging: a sociological approach*, Milton Keynes: OUP.

Cabinet Office. (1998). *Better Government for Older People.* London: H.M.S.O.

Chau, W.F. (1995). 'Experts, networks and inscriptions in the fabrication of accounting images', *Accounting Organisations and Society*, 20, 2/3, 111-145.

Chen, S and Powell, JL(Eds.) (2011) *Aging in Perspective and the Case of China*. Nova Science: New York.

China Daily (2004), "Tailoring Health Care Policies for the Elderly", April 15th 2004.

Chudacoff, H. (1989). *How Old are You?* New Jersey: Princeton.

Cohen, S. (1985). *Visions of Social Control*. Cambridge: Polity.

Cole, T., Van Tassel, D. & Kastenbaum, R. (1992). *Handbook of the Humanities and Aging*. New York: Springer.

Conrad, S. (1992) 'Old Age in the Modern and Postmodern Western World' in Cole, T (Ed.) *Handbook of the Humanities and Aging,* New York: Springer

Cook, I.G. and Murray, G. (2001), *China's Third Revolution: Tensions in the Transition to Post-Communism,* Curzon: London.

Cook, I.G. and Powell, J.L. (2007) *New Perspectives on China and Aging*. Nova Science: New York.

Cousins, M. & Hussain, A. (1984). *'Michel Foucault'*. London: Macmillan.

Cumming, E. and Henry, W. (1961). *Growing Old: The process of disengagement*. New York: Basic Books.

Davidson, A. (1986). 'Archaeology, Genealogy, Ethics' in Hoy, D (Ed.). *Foucault: a critical reader*. Oxford: Basil Blackwell.

Dittman-Kolhi. F. (1991). The construction of meaning in old age. *Aging & Society*. 10, 279-94.

Du, P. and Tu, P. (2000) 'Population Aging and Old Age Security', chapter in in Peng X. and Guo Z. (eds), *The Changing Population of China*, Oxford: Blackwell, pp. 77-90.

Epstein, H. (2001) 'Time of Indifference', *New York Review of Books*, April 12, pp.33- 38

Estes, C. (1979). *The Aging Enterprise*. San Francisco: Jossey Bass.

Estes, C. and Associates (2001) *Social Policy and Aging*. Sage: Thousand Oaks.

Estes, C. Swan, J. and Gerard, L. (1982). 'Dominant and competing paradigms in gerontology: Towards a political economy of aging', *Aging and Society*, 12: 151-64.

Estes, C., Biggs, S and Phillipson, C. (2003) *Social Theory, Social Policy and Aging*. Open University Press: Milton Keynes.

Featherstone, M. & Wernick, A. (1995). *Images of Aging*. London: Routledge.

Featherstone, M. and Hepworth, M. (1993). 'Images in Aging' in Bond, J. Coleman, P. and Peace, S. *Aging in Society*. London: Sage

Federal Reserve Bank of Kansas City (2004) *Global Demographic Change: Economic Impacts and Policy Challenges. Symposium proceedings*. August 26–28, 2004. Available at: http://www.kc.frb.org/Publicat/sympos/2004/sym04prg.htm

Fortes, M. (1984). 'Age, Generation and Social Structure' in Kertzer, D. and Keith, J. (Eds.) *Age and Anthropological Theory*. London: Cornell University Press

Foucault, M. (1972), *The Archeology of Knowledge,* Tavistock: London.

Frank, A.W. (1998), Stories of Illness as Care of the Self: A Foucauldian Dialogue, *Health*, Vol. 2, No. 3, pp. 329-48.

Gavrilov, L and Gavrilova, N (1991) *The Biology of Life Span: A Quantitative Approach.* New York: Harwood Academic Publisher.

George, L. (1995). 'The Last Half-Century of Aging Research - and Thoughts for the Future', *Journal of Gerontology: Social Sciences*, 50, (B), 1, 1-3.

Giddens, A (1993) *Sociology*. Cambridge Polity Press.

Gilleard, C and Higgs, P. (2001). *Cultures of Aging*. London: Prentice Hall.

Government of China (2002), *China Statistical Yearbook 2002*, China Statistics Press: Beijing.

Gruber J, and Wise DA, (eds.) (1999) *Social Security and Retirement around the World*. Chicago, IL: University of Chicago Press, 1999.

Gruber J, and Wise DA, (eds.) (2004) *Social Security Programs and Retirement around the World. Micro Estimation.* Chicago, IL: University of Chicago Press.

Hadley, R. and Clough, R. (1996), *Care in Chaos*, Continuum International: New York.

Hansson, A. (1996), *Chinese Outcasts: Discrimination and Emancipation in Late Imperial China*, E.J. Brill: Leiden, The Netherlands.

Harper, S. (1994), China's Population: Prospects and Policies, in D. Dwyer (ed.), *China: The Next Decades,* Longman Scientific and Technical: Harlow, UK.

Heidegger, M. (1971), *Poetry, Language, Thought,* Harper and Row: New York.

Help the Age International (2000) *The Mark of a Noble Society.* London: HelpAge International.

Hermalin A, (ed.) (2002) *The Well-Being of the Elderly in Asia: A Four-Country Comparative Study*. Ann Arbor, MI: University of Michigan Press

Holtzman, R.A. (1997) *A World Bank Perspective on Pension Reform.* Chapter prepared for the joint ILO-OECD Workshop on the Development and Reform of Pension Schemes, Paris, December.

International Monetary Fund. *The Economics of Demographics. Finance and Development. September* (2006);43(3). Available at: http://www.imf.org/external/pubs/ft/fandd/2006/09/

Jones, C. (1983). *State Social Work and the Working Class.* London: Macmillan.

Jordan, B. (1996). *A Theory of Poverty and Social Exclusion.* Cambridge: Polity.

Kastenbaum, R. (1993). Encrusted elders. In Cole, T. (Ed). *Voices & Visions of Aging*. New York: Springer.

Katz, S. (1996), *Disciplining Old Age: The Formation of Gerontological Knowledge,* The University Press of Virginia: Charlottesville, VA.

Katz, S. (1998). 'Lifecourse, lifestyle and postmodern culture: corporate representations of later life, chapter presented at *Restructuring Work and the Lifecourse: An International Symposium*, Institute of Human Development, University of Toronto.

Katz, S. (2000), Busy Bodies: Activity, Aging and the Management of Everyday Life, *Journal of Aging Studies*, Vol. 14, No. 2, pp. 135-52.

Kim, S., and Lee, J-W., (2007), "Demographic changes, saving and current account in East Asia", *Asian Economic Chapters*, 6(2)

Kinsella K, and Velkoff VA. (2001) *An Aging World: 2001*. Washington, DC: National Institute on Aging and U.S. Census Bureau

Krug, E.G. (2002), *World Report on Violence and Health*, World Health Organization: Geneva, Switzerland.

Laws, G. (1995). Embodiment and emplacement: identities, representation and landscape in Sun City retirement communities. *International Journal of Aging & Human Development*. 40.4: 253-280.

Levick, P. (1992). 'The Janus face of community care legislation: An opportunity for Radical Opportunities' in *Critical Social Policy*, Issue 34, Summer 1992, pp.76-81.

Liu, J. and Lin, F. (2000), *Long-Term Effect of China's Family Planning.*

Longino, C. F. (1994). 'Pressure from our aging population will broaden our understanding of medicine'. *Academic Medicine, 72* (10), 841-847.

Lopez AD, Mathers CD, Ezzati M, Jamison DT, and Murray CJL, (eds.) (2006) *Global Burden of Disease and Risk Factors. Washington*, DC: The World Bank Group

Lubove, R. (1966). 'Social work and the Life of the Poor', *The Nation*, 23 May, pp.609-11.

Manton KG and Gu X. (2001) *Changes in the prevalence of chronic disability in the United States black and nonblack population above age 65 from 1982 to 1999*. Proceedings of the National Academy of Sciences 98;6354-6359.

May, T. and Powell, J.L. (2008) *Situating Social Theory 2*. McGraw Hill: Maidenhead

Moody, H. (1998). *Aging, Concepts and Controversies*. Thousand Oaks: Pie Forge Press.

Murray, G. (1998), *China: The Next Superpower*, China Library: London.

Murray, G. (2004), *China's Population Control Policy: A Socio-Economic Reassessment*, PhD thesis, Liverpool John Moores University: Liverpool, UK.

Nettleton, H., Biggs, S., Bernard, M, Knigston, P. (In press) Focus groups and older people. *Qualitative Health Research.*

Neugarten, D. (Ed.). (1996). *The Meanings of Age*. Chicago: University of Chicago Press.

Organisation for Economic Cooperation and Development (OECD) Directorate for Employment, Labour and Social Affairs. (2007) *Disability Trends among Elderly People: Re-Assessing the Evidence in 12 OECD Countries* (Interim Report). Paris, France: OECD

Patel, N. (1990.) *A Race Against Time*. London: Runnymede Trust.

Phillipson, C (1988). 'Challenging Dependency: Towards a new Social Work with Older People' in Langan, M. and Lee, P. (Eds.) *Radical Social Work Today*. London:Unwin Hyman.

Phillipson, C. (1982). *Capitalism and the Construction of Old Age*. London: Macmillan.

Phillipson, C. (1998). *Reconstructing Old Age*. London: Sage.

Phillipson, C. and Biggs, S. (1998). 'Modernity and Identity: themes and perspectives in the study of older adults, *Journal of Aging and Identity*, 3, 1, 11-23.

Powell, J. (1999), 'The Importance of a 'Critical' Sociology of Old Age', *Social Science Chapter Publisher*, 3, 1.

Powell, J. (2001), Theorizing Gerontology: The Case of Old Age, Professional Power and Social Policy in the United Kingdom, *Journal of Aging and Identity*, Vol. 6, No. 3, pp.117-35.

Powell, J.L. (2005) *Social Theory and Aging*. Rowman and Littlefield: Lanham.

Powell, J.L. (2009) 'Social Theory, Aging, and Health and Welfare Professionals: A Foucauldian "Toolkit", *Journal of Applied Gerontology,* 28, 6, 669-682

Powell, JL (2011) *Aging, Theory and Globalization*. Nova Science: NY.

Powell, J.L. and Biggs, S (2000). 'Managing Old Age: The Disciplinary Web of Power, Surveillance and Normalisation', *Journal of Aging and Identity*, 5, 1, 3-13.

Powell, J. and Biggs, S. (2000), Managing Old Age: The Disciplinary Web of Power, Surveillance and Normalisation, *Journal of Aging and Identity*, Vol. 5, No. 1, pp. 3-13.

Powell, J. and Biggs, S. (2001 forthcoming) 'Bio-ethics and Technologies of the Self: Understanding Aging', *Journal of Medical Humanities*

Powell, J. and Cook, I.G. (2000), "A Tiger Behind and Coming Up Fast": Governmentality and the Politics of Population Control in China, J*ournal of Aging and Identity*, Vol. 5, No. 2, pp. 79-90.

Powell, JL and Cook. I (2010) *Aging in Asia*. Nova Science: NY

Rabinow, P. (Ed.). (1984). *The Foucault Reader*. Paris: Peregrine.

Satyamurti, C. (1974). 'Women's Occupation and Social Change: the Case of Social Work, chapter presented to 1974 *British Sociological Association's Annual Conference*.

Shilling, C. (1993). *The Body and Identity*. London: Routledge.

Shumway, D. (1989). *Michel Foucault*, Charlottesville: University Press of Virginia.

Smart, B. (1985). *Michel Foucault*. London: Routledge.

Townsend, P. (1981). 'The Structured Dependency of the Elderly: A Creation of Social Policy in the Twentieth Century' in *Aging and Society,* Vol.1, No.1, pp.5-28.

Toyota, M. (2006), Aging and transnational householding: Japanese retirees in Southeast Asia, *International Development Planning Review*, 28, 4: 515-531.

Turner, B. (1989). 'Aging, Status Politics and Sociological Theory' in *British Journal of Sociology*, 40: 588-606.

Turner, B. (1995). Aging and identity. In Featherstone, M. & Wernick, A. *Images of Aging*. London: Routledge.

United Nations Department of Economic and Social Affair (2002) *Population Division. World Population Aging 1950–2050.* New York: United Nations

Veyne, P. (1980). *Foucault revolutionne l'histoire*. Paris.

Wahidin, A and Powell, J (2001) 'The Loss of Aging Identity: Social Theory, Old Age and the Power of Special Hospitals', *Journal of Aging & Identity*, 6, 1, 31-49.

Walker, A and Naeghele, G (2000) *The Politics of Aging in Europe*. OUP: Milton Keynes

Walker, A 'Towards a Political Economy of Aging', *Aging & Society*, 1, 2, 23-44.

Wilson, G. (1991). 'Models of aging and their relation to policy formulation', *Policy & Politics*, 19, 1, 37-47.

Woolfe, R. and Biggs, S. (1997). Counselling older adults. *Counselling Psychology Quarterly*. 10.2: 189-94.

World Bank (1994), *Averting the Old-Age Crisis*, Oxford University Press: Oxford, UK.

Zheng, L. (2004), China Faces Elderly Dilemma, *China Daily*, August 21.

INDEX

D

E

S